Breathing with the Climate Crisis offers a new hopeful narrative, a different perspective that could unleash the courage to act. Young people and farmers from the East, South, North and West asked at a world biodynamic conference, *'How can we find our own breath? Do we need more facts? More head? More heart? Feeling? Poetry?'*

Beginning your own inquiry process, ask yourself questions. Think further. Also, with your heart and hands…

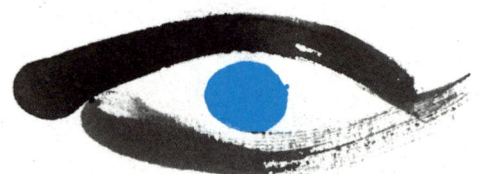

Someone asks.
Is my body the earth?
I am the earth.
The earth is me.
My body is the earth.
The earth is my body.
But I am not just a body.
And not only the earth.
Part of me is a guest.
We are guests.
But also at home.
I breathe.
The earth breathes.
The earth lives.
I want to live on the earth.

Breathing with the Climate Crisis

A new beginning

This is for you. You can answer.
Or listen. Pass it on or ignore it.
It is up to you. It is your decision.

We need air.
We need breath.

We need a new horizon. A new narrative. **A new perspective.**
We know. Almost everything. We know about climate change and the loss of biodiversity. Animals are dying, plants withering.
We know about us human beings who do not listen to each other. We know about the challenges of our times. We know so much.

Theoretically we know what we would have to do. Could do. Should do. Must do. But we rarely do anything. We hear shouts, we fight an invisible enemy. We fight something. We fight each other. We fight climate change. We fight the changes of our times.
We fight the destruction of the earth.
We have a feeling of suffocating. In what we must do, could do, should do, in what we know. In the challenges of our times.

How can we draw breath nowadays? How can we breathe when our footprint only does more damage? How can we breathe, when our action contributes to more CO_2, less biodiversity and greater injustice in the world?

I am seeking. New perspectives. A new viewpoint. Of our role as human beings. Fellow human beings. On our earth. Our relationship with each other.

We can do it. You. I. The earth. Together.

Together we want to make a positive footprint. We want to breathe. We want to act.

It needs you. Some things we can do alone. But many only together. Our times are asking for a change. All the way down to the roots. A radical change, a quick transformation.
Get started!

We are the earth.

One chapter is coming to an end. We have reached a full stop. The new chapter will follow. The new story. We are those who write and those who are written about. **Just start writing!**

*Our mother earth has hardened
Through pain.
Our mission is to
spiritualise her again,
by reworking her through
the power of our hands
into a spirit-filled work of art.*

Rudolf Steiner, over 100 years ago

The personal designations in this text apply to all genders.

1 Your earth 14

2 You human being. You as co-creator. 26

3 From real life – every person is a farmer 36

4 Wanting a global society?! 48

Further reading 52

Background 53

Short biographies of the authors 56

Publishing information 58

Let us open up a new horizon!

Your earth

It has already been said. From indigenous peoples all the way to Johann Wolfgang von Goethe. All over the world.

The earth is a living organism. A being. A counterpart. With her own rhythms. With her own development, a biography. With her own task in the great fabric of the cosmos.

Just assume for a moment that we had a different point of view of our earth. A different one from the one we have nowadays.

A place where plants and animals have their kingdom. A place where man has arrived. A place where all creatures live in the breathing rhythm of this earth. Can you experience the rhythm?

The earth sees, sees you. Bears you. Feels you. Hears you breathing. She is you. You are she.

Yes, a change of perspective!
What would change? How would our relationship to the earth change? Would our actions change? Yes, our actions.

Let us accept the thought experiment of the earth

as an organism. She is not a lifeless object, no. Consisting only of minerals and resources. She is not a broken machine with replaceable parts. Your earth is part of the cosmos in our solar system. The place where you are currently sitting and reading these words.

My actions in one place also affect your place. They can be experienced directly. Sometimes it takes time to become aware of it, but it can nevertheless be experienced. My action influences another place far away from me.

Global thinking enters the local action that I take. And vice versa. Seeing and learning complexity. Feeling empathy, experiencing the earth as an organism. Life. Allowing life. Enlivening her. Helping to create her.

A great task. Images help as a bridge to find new perspectives.

We are human beings. Various organs act in us. Their interplay enables us to breathe, live and develop free will. If one of the organs is injured, then the interaction of the whole organism is impaired. We get hot. Get a fever. We feel a change of climate in our own bodies.

Our immune system is often able to help. But if important organs, the lungs or the heart, are damaged, it is difficult. Difficult to act effectively and healthily. Difficult to go on.

There are limits when particular values are exceeded. Too little blood or too many harmful substances in our bodies. Then everything collapses. This can lead to the death of the whole organism. Death can occur.

But you are nevertheless more than this. More than the sum of your separate parts. The skin encloses an independent being. Your being. Carried by its own rhythm. Breathing, digestion. Waking and sleeping phases. Are you? Are you awake?

Let us apply this picture to our earth. She has a skin. She has organs: seas, rivers, landscapes, fields, plants, pastures and animals. Mountains and minerals. Our earth, the blue planet, has a heart formed of oceans. The rainforest, the Amazon basin, is not called the lungs of the earth for no reason. The soil with its breathing humus layer is something like a diaphragm.

Yes, we know this. We have already heard about

it. But have we really thought about this image, internalised it?

> *The earth is a living organism, not a broken machine that can be repaired.*

Let us go more deeply into the ideas about the earth organism.

Our soils are the bearers of life. They are one of the most essential organs of our earth.
They are a basis. For our food. Minerals for our technology. Precious stones and fossil fuels as fuel for our old society.

We are destroying this basis. Eating her up. Eating up our own existence. The ground under our feet.

Destruction is no longer an option.

With our actions, we are pushing the heartbeat of the earth towards zero. Finished. Over. An end? You are the heartbeat of the earth. Run! Run! Movement.

Let us look further at carbon.

Everything on this planet has a particular function.
Each organ has a meaning.
The role of carbon for our earth. Who knows what it is? Who is looking for it?

Everything organic on our planet consists of carbon to a certain degree. **Carbon combines. A living and creative task.**

Right from the outset, plants devote themselves to the task of absorbing carbon and enabling the animals and people on this planet to breathe. Fossil fuels, petroleum and coal. Once upon a time they were living plants.

Irony? Something that once gave us air to breathe, in its combusted form now takes the air away.

The balance between the giving of plants and the taking of humans. Where is it? Is it a strong and healthy breath?

Carbon has also created a connection in universal language. Measured and expressed in positive and negative CO_2 equivalents. These are universally valid and clearly, technically defined. They now function as a currency, as the basis for decision making, but also as an expression of acute threat.

Carbon has become a language of global understanding. It enables a global debate. We have joined in. Perhaps carbon is not just something bad. But only at too high a concentration in the wrong place. What are the roles of carbon? Destroying? Creating? Killing? Giving birth? Breathing in? Breathing out? Calm.

What would happen, for example, if we humans were to recognise the function of carbon in the earth's organism as an element of creation and vitality?

Carbon as the approach to a solution. When farmers actively bind it in hedges and soil. They create. They nurture!

Could carbon become the basis for a renewed, healthy breathing? Does the breath need a different rhythm?

Breathing is rhythm. Rhythm is breathing.

Learning rhythm. Learning to understand. Every human, every organism has its own rhythm. Is the organism looking for relationship? Integrated into the cosmic fabric. Relationship to sun, moon, planets and stars. You are cosmos.

Our heart beats differently when we are healthy than when we have to fight a feverish flu. Our sleeping and waking rhythm are affected by our activities during the day, from what we have given ourselves as spiritual and physical nourishment and from our general stress level.
The heart beats cosmos.

Have we as human beings forced our own, new and not really suitable rhythm onto the being of the earth?
Have we knocked the earth off balance? The natural phases of breathing in, breathing space, breathing out and breathing space have been interrupted. We take. Without a break. Without the opportunity of allowing the earth to breathe in again.

Run! The earth needs you. With a break. Breathe in, breathe out. Allow the inbreath. Allow the outbreath. The earth.

If we consider rhythm, then time belongs here too. What is now? What is tomorrow? What is the future?

We as people have lost ourselves in the past in short-term action. Short-term action as our main business. Faster! Further! Higher! More! I'm falling. Stop.

We have also lost our rhythm. Short-sightedness.
We need farsightedness.

I want to be a healthy rhythm. A healthy view needs time.

Breathing. Wanting instead of necessity. Wanting to breathe.

The earth bears your action. Diagnosis alone does not bring about change. We belong to a generation that has never had so much access to freely available knowledge. We are complete. Know everything.
But still, action seems so difficult.

We go by car. We fly. We eat. We buy. We throw away. We create. We destroy.
We don't hear. We want to live! How?

Real consistency in the climate crisis is complex. It is sometimes inconvenient and can require overcoming our own comfort.

Breathing the environmental crisis. Sensing. Getting to know. Looking away. Not looking away.

Looking.

What viewpoint could help us to find a new way of treating our earth? **The image of the earth as an organism can help us to connect to the earth.** To develop empathy and to begin positive action.
We believe that it is not in our nature to want to *hurt* other living creatures, hit them or even maltreat them. Our being, your being, my being. Does not hit. Does not steal. Does not kill.
But if we do, we do not feel *good* about it. Our own existence becomes a great burden of debt. We can no longer breathe freely. Don't get enough air.

The earth as a real being gives us a counterpart.
It can help to live up to our responsibility. Helping the earth, finding her healthy form instead of injuring her with chemicals and exploitation. Influencing the earth in a positive way.

Now the question arises as to what we do if we have injured the earth as a living being too greatly. Do you run away? Into space? Into the new, digital world?
To live there in the future?

Do we as humanity really want to leave an injured being behind?
I don't. Do you?

We love the earth!

# You human being.
You as co-creator.

**Every part counts. All count.
You. Earth. I. We.**

Every organ in the organism is relevant. Is a part of the interaction in the whole picture of our earth.
So the human being is also an important part of this.
Hard work happens on the earth. Development. Development is needed. **We want to develop ourselves.** Human beings seek new methods, technical solutions and more.

The MORE is a credo. More of everything.

While the resources are dwindling, the earth ruined. She is overheated. Now it's about repairing the broken machine of the earth. The aim of this is to carry on with our own living standard.

We see a machine. An overheated machine that needs to be cooled down. Not tomorrow but now. Actually yesterday. She needs to be cold. Cold?

We are looking for replacement parts for the earth. Getting a supply of coolant. In order to capture our own poisonous emissions and make them disappear under the earth again. Where? There. Do we want to see the earth like this? Tell each other that's the way it is?

This brings up questions for us: do we human beings want to stay on this earth? Will you stay? Do we have enough opportunities here to develop ourselves? Do we want to remain a part of the earth? Or do we break off the partnership and go another way? Leave the earth alone and go away?

Departing Humanity. Will we leave the earth?

A new genius is driving us away. Away from the earth. Into space. And further, further! As far as technology will allow.

Billions are being invested in working on enabling human beings to leave the earth. For human beings to continue their existence independently of the earth. To free human beings, allow them to disappear.

Billions are being invested to launch us into space. Launch! 3...2...1...LAUNCH.

For years, Hollywood and science fiction novels have been presenting ideas of a future where the earth is destroyed. In these scenarios, the human being disappears. Flies away from the earth. We are resilient. But are we this resilient?

We are resilient. But space is as well. What about the earth? But the earth does not want to do without. Do without you. Without me. Without us. Without a future. She is on her path. Do not destroy her. Help her!

What we possess as human weaknesses could in future be technically replaced by machine parts. Transhumans. The human being is disappearing.

People, who have become robots in part. Survive in extreme situations. Survive. They ought to survive. Conquer the process of disappearing. They should land on the earth, arrive. We are to become transhuman. Become the earth.

They know neither illness nor death. The human robots. Death and birth become one. Standstill in space. In the universe. The universe comes to a halt, holds its breath.

And so the riddle of the creation of humans and the earth is solved. **What no one has achieved,**

humans achieve. Make themselves disappear. From head to toe. Body and soul. With one and all.

Is this our task? Is this your task, here and now?

Not me. Not us. Not you?
We want to live on the earth!

Let's first examine the consequences of our own actions.
The beginning of the Anthropocene, the era of human beings. Clever heads fix a date. When the first atom bomb experimentally exploded in the desert of New Mexico. Our history starts with the greatest destructive force.

Man appears as a tyrant. As the destroyer of his fellow humans and the earth. The picture of tyrannical humans separated from nature leads to our wish to be "less human". We have the feeling that human beings must leave, in order to protect nature and the earth from us. We as human beings get the feeling of having no justification, because we destroy, kill, heat the planet.

Do we have to leave? Disappear? To protect the earth from us?

Yes, in a certain way we need to accept that our decisions and views in the past did not serve either us or nature. We have created a technical, artificial separation from the earth that has not helped us.

We have created a separation from the earth. We were free. So free, that we wanted to fly away. Fly away into space. Away from our task. I question this. We question this. Do you question this?

We saw the earth as a machine. A machine that overheats. We applied a reduced, technical, even mechanical point of view to her. We thought ourselves bigger, more independent and freer than we actually are. The earth machine is running hot. Where is the switch? In me, in you. Switch over, to the place where I breathe, you breathe. Where we breathe. Breathe with the earth.
At a personal level, we recognise that knowledge of the climate alone does not lead to positive actions.

We have failed. Can we honestly admit this?

The human being is inhuman. But the human being is not inhuman. Does not need to be so. You are free. We are free. We are many.

We see what was. See what can become.
Courageously take a new breath.

The destructive force is ONE part of the human being. But we have many facets.
We can decide. Every day. Anew.

We can decide in favour of a positive creative force. In complete freedom.

For us as living beings with responsibility and tasks for the development of the earth. As co-creators of the earth. Instead of having to protect her from us. **For more human being.** For more human being in their full, healthy strength.
More creative and developmental powers instead of destruction. More of an anthropocentric view instead of rationalising humans away.
Is this perhaps the task of the 21st century for us humans? Transforming the killing power of human beings into creative power for a new world.

The climate crisis acquires a new dimension: breathing through the climate crisis? What is the task of the human being in these times?
Environmental climate crisis. Social climate crisis. Cultural climate crisis. Are these symptoms of current events? Our current events. As a mirror.

Our mirror. My mirror. Your mirror. Dimensions for transformation.

Moments of hope. I want to grant strength. To you. To us.

The human beings who sense and love the earth as a living organism, who want to co-create in community, require a new understanding and definition of the Anthropocene. They advocate hope. Hope that humanity wants to live with the earth and can live with the earth.

The earth's health is your health, my health. Sustainability and resilience speak to you, to me. They speak to your health, my health, our health.

In search of someone who can make a relationship to the destiny of the earth. In search of someone who can feel responsibility for the destiny of the earth. The destiny of the earth. Your earth. And yet you are also a guest.

Not 'Departing Humanity',
but a humanity that wants to approach the earth as a partner.
We want to live on the earth!

A shared world and earth. Simply protecting and

preserving nature are past. A healthy relationship with the earth. A fair, respectful, attentive partnership.
A viewpoint based on the idea that the earth is in dialogue with its mineral nature, its plant nature, its animal nature, with the human as a creative being.

Can we also give, give more than take? How can the human being actually become a co-creator? Daily life as the driving force. As in every relationship. As the main activity for recovery.

Is this a task for culture? A task for nature? A task for culture and nature? Can nature and culture be reconciled?

How can culture integrate again into nature?
Forge a close relationship?
Is this the task of human beings?

The relationship to fellow humans, to the other. The search for a third entity that demands that the dialogue becomes a trialogue. Increases and progresses.

Where do I see a transformative society?

The human being is a co-creator instead of a destroyer of the earth.

From real life – every person is a farmer

What would it be like if we were to acquire a radically different view of our earth?

If the earth were an independent organism and human beings had the opportunity of entering into a responsible partnership with her? What would change in our society?

What would it be like if this view of things had already been put into practice?

Throughout the world. Here and now. Every day.

How would it be if this was already happening and more people could find out about it?

It does exist: biodynamic agriculture. This is an example. An example of an earth-organism-agriculture. An example of an earth-cosmos-agriculture. An example of a human-earth-sun-agriculture.

There are other examples. We can learn from each other. For the earth. From the earth. From the sun

in the earth. For agriculture. But also in general for life on earth. For our life on our earth.
Every person is a farmer. You too? You too! Directly or indirectly we all influence the soil. One, the farmer, greets the sun as it rises above the horizon across the pasture, the other eats the fruits of the earth. The sun-earth, heavy with fruit. You have an influence. On what happens there. On what happens here. On the earth. Every day anew.

Biodynamic agriculture produces food and biodiversity at the same time. It creates relationship with fellow human beings, the earth and the laws of day and night, sun and moon, earth and cosmos. It works with the rhythm of our solar system.

For example, in biodynamic agriculture the idea of the organism is successfully lived in a very practical sense on many farms around the world. This is called the "farm organism".
The farm is a living being with a closed cycle of substances. Feed – animals – compost – soil – crops. Clear and effective. Rooted in local conditions, but with a gaze towards the stars.

The farm as a living being with various elements

and organs. The human being is actively involved as a creative individual, as a co-creator of the earth.

The farmer harmonises their decisions with the surrounding environment.
This requires finely-tuned observation skills and regular practice in the perception of oneself and nature. Practise. Again and again. Observe. The landscape, the atmosphere. Earth and topography, water and humidity, air and wind, warmth and light. The historical development of the landscape and its diverse life are taken into account. This results in a unique form of agriculture at every individual place.
We can smell this, taste it. Fully. You. Smelling and tasting. The unique, individual quality of the products of this place: the Genius loci. The sum is more than its separate parts.

What can we learn from these farmers?

That we too can first observe, perceive and get to know nature and our earth before we make decisions. **We can learn to harmonise our decisions with the needs of the earth.** To enter into a dialogue.

This may mean that we sometimes move outside our comfort zone. If a calf is born in the middle of the night and the cow needs some help, you as the person responsible for your farm do not consider for long whether this is pleasant for you. You get going. **You take on responsibility for** the living earth, in the place where you live and work. But you also take on responsibility for the things you eat, the clothes you wear, use and buy.

Letting go off what is familiar can be an enrichment. **Wanting to learn new things.** About yourself. About your surroundings. About your earth.

You are allowed to be an inquisitive discoverer. Discoverer of your needs. Your boundaries. The boundaries of your counterpart, your earth. But also the opportunities. Your opportunities. The opportunities of a partnership.

Our ability to open ourselves can carry us through. Awakening: what are all the possibilities? All this is possible!
It can awaken the joy to enter into partnership with the earth. Like a new friendship, when you have just got to know each other.

New ideas can arise. New directions. Most biodynamic farms are known to be pioneers in their regions. Established from the impulse to take up farming in a new way and tread new paths. They live intellectual curiosity and eagerness to try out something new. **How much a partnership with the earth can achieve!**

It is not about having to change 'everything', 'now', 'immediately'. Obviously, we need a radical transformation. A transformation that goes right to the roots. And it is needed today rather than tomorrow. But time pressure should not be allowed to lead to paralysis so that we finally stop moving altogether. We lose our breath and our joy over the transformation and new partnership.

Some biodynamic farms go for the long-term. They transform the different parts of the farm thoughtfully, one part at a time. They design the individual areas conscientiously so that together they produce a complete picture.
You will seldom find a farm where everything is done at once: trying out new arable crops, extending the sheds for the animals, installing alternative electricity supplies, developing new marketing channels. A farm usually consists of continuous, small and specific adjustments. Bit by

bit. Day by day. **Always deciding in favour of the partnership instead of against it.** For instead of against. Wanting a transformation. This makes the farmer into a carer for the climate. Step by step with the whole in view. And in their heart. The whole in view and in their heart.

This also applies to you. **You can establish new habits in your life, step by step.** You can be a farmer, a carer for the climate, even a carer for the earth. In fact, you already are. Wanting to look again at all the daily actions. Look properly at everything that your action affects. Ask, what it is, where it comes from, where it goes to. Ask. Look. Act.

Biodynamic agriculture avoids the use of chemicals and genetically modified seed. This supports biodiversity, soils, water resources, the atmosphere and human health. This will lead to the future. Will develop the future. For the earth. For the soils. For biodiversity. For future generations. This cannot be done without us. Not without us human beings.

Financially it works. Usually. Good products can have a fair price. In those societies which have enough money. Less is more. Less food waste, more quality. A choice that everyone has to make. Your choice? Your choice!

The use of chemicals and genetically modified seed is expensive for the farmers. In India, in Tanzania, in Peru. Less expensive, with a light touch. It's possible. Do you know them? Your farmers in India, Tanzania, Peru? They who produce your curry, your coffee, your cotton with such a light touch?

Expensive also for the earth, humanity and our future. It is not an option to want to produce so much in the short term, with the knowledge and awareness that this kills the soil in the long term.

The soil can live. Biodynamically. Breathe. It is a respiratory organ in the farm organism. Comparable to the diaphragm in our body. Breathing in the climate crisis. The soil can do this too. With humus. With soil life. With roots. With earthworms. With structure. As a dialogue between the earth and the cosmos. Between depths and heights. This can be culture. Agriculture. By human beings. Human! Humanum! Humus!

In the soil! With carbon. Not in the air as CO_2 or as nitrous oxide (N_2O) or methane (CH_4). Better to keep these substances in the life cycle. Keep them in life, in the soil, instead of emitting them. This works better without chemical fertilisers. As was shown by the DOK study 40 years ago, scientifically and with plots of land with different treatments.
D = dynamic. O = organic. K = conventional (FIBL's longitudinal 40 year DOK study).
Without chemical fertiliser: 20 % less yield, 50 % more energy efficiency, 30 % less nitrous oxide emissions. **More life. In the soil. It's possible.**

Fertile soil is built up from applications of compost. So compost becomes the heart of a farm. It closes the cycle and contains the past and future of the place. It is the new gold.

Biodynamic farms also teach us that it only works when we pull together. They are masters in establishing Community Supported Agriculture (CSA) and social forms of farming. Building bridges with growing agreements. With short supply chains. Between town and country. For a new consumer awareness of the treasure that is produced on the fields and in the animal sheds. Doing business together. Associative economics.

With profit. For all. For the farmer. For you. For the earth.

In other words: talk to others about your experience! Exchange ideas! Build networks and get involved with others for partnerships with your earth! Share your success and failure!

Yes, it might be difficult to talk about it, but it enables you to breathe. To draw breath. To develop new ideas with others on how to do it differently and perhaps better in future.

But what does all this do for you? What can you learn from it?

Cooperation between the earth and human beings is good for us. It is part of our nature. This is something else that can be learned from biodynamic agriculture. The people who run their farms with love and affection are usually very content. They work from an inner strength. They work for love of the thing. The connection to the soil, the plants, animals, landscape, earth and sun have a healing effect.

Cooperation with the earth is open to all. Also to you. If you want. You are free. It is a matter of breathing. How can you breathe in the climate crisis?

Look inside yourself to see which decisions allow you to breathe.

Engage in practical work.
Let the earth feel your conscious footprint.
A positive one this time. She needs us.

Every person is a farmer.

Illusion? Vision!

Wanting a global society?!

How can I breathe? Socially, culturally, environmentally and spiritually? How can we breathe?

Can a new, healthy breath arise from a newly wanted global society? A society that sees, affirms and welcomes the earth, cultural plurality and every individual? A global society that can affirm the whole in a healing way?

Does it exist? In our thoughts? In our hearts? In you?

Illusion? Vision!

The call in you, deeply linked to you, to all, to the earth. Will something be born?
Something wants to start to breathe. **Breathing with the climate crisis.**
It wants to be born. Out of love. To come to the earth. The earth. Love. It wants to become love. It wants to become the sun. It wants to become light-warmth.

Humans, earth, sun.

The You made the I possible. The I makes the You possible. Because of I and You there is the world. There is the earth. The earth is our third partner. Let us open our hearts to her, our dialogue, include her, listen to her.
She has spoken, the earth. It is your turn again. And mine. It is our turn again. The earth has brought us together. Has created us. Mother earth. She is the substance of our destiny. The earth as a potential place of peace.

But now, stop the destruction! Overcome the old story, the old viewpoint!
Do not get angry, do not be ashamed of the past!
We wish to go on into the future together.
Into the now. Creating a global society.
You are a protagonist.

Make a contribution instead of THE solution!

Associate so that human beings can make a cultural contribution to the earth.
A contribution out of the full, healthy and creative power of each individual person. **From your place. Locally. For the whole. Globally.**

> *What do you need?*
> *Are you ready?*

My life
the earth
environment
environmental degradation
climate crisis
CO_2
reductionism
numbers
machines
death
zero point
death
machines
numbers
reductionism
CO_2
climate crisis
environmental degradation
environment
life
the earth as a living organism
the human being as a co-creator of the earth.

AND
NOW
YOU

Further reading

Your earth
- Rudolf Steiner, *The philosophy of freedom: The basis for a modern world conception,* Rudolf Steiner Press, London, UK, 2011.

You human being. You as co-creator
- Charles Eisenstein, *Climate: a new story,* North Atlantic Books, US, 2018.
- Dan McKanan, *Eco-Alchemy: Anthroposophy and the History and Future of Environmentalism,* University of California Press, USA, 2017.
- Wulf, Andrea, *The Invention of Nature: Alexander von Humboldt's New World,* First American Edition, Vintage Books, 2015.
- Von Weizsäcker, Ernst Ulrich, and Anders Wijkman, *Come On!,* Springer New York, 2018.

From real life – every person is a farmer
- Biodynamic Federation Demeter International (BFDI): https://demeter.net/ (20.09.2022)
- Research Institute of organic agriculture (FiBL), *The DOK Experiment: Long-term study on bio-dynamic, bio-organic and conventional farming systems*, https://www.fibl.org/en/themes/projectdatabase/projectitem/project/404 (20.09.2022)
- Section for Agriculture, *Basics of biodynamic agriculture*, https://www.sektion-landwirtschaft.org/en/basics/biodynamic-agriculture (20.09.2022)

- Rudolf Steiner, *Agriculture Course: the birth of the biodynamic method*, Rudolf Steiner Press, London, UK, 2004.
- Maria Thun, *Gardening for Life - the biodynamic way*, Hawthorn Press, Stroud, UK, 2016.
- Marina O'Connell, *Designing Regenerative Food Systems: and why we need them now*, Hawthorn Press, Stroud, UK 2022.

Wanting a global society?!
- Rudolf Steiner, *World Economy*, Rudolf Steiner Press, London, UK, 1990

Biodynamic Association UK
https://www.biodynamic.org.uk/
office@biodynamic.org.uk

Biodynamic Association USA
https://www.biodynamics.com/
info@biodynamics.com

Biodynamic Agriculture Australia
https://biodynamics.net.au/
bdoffice@biodynamics.net.au

Biodynamic Association New Zealand
https://biodynamic.org.nz/
info@biodynamic.org.nz

Background

This appeal arose from a conversation. Several people were gathered together. Different generations. Different vocations.
They were planning a conference on the topic of *Breathing with the Climate Crisis*. A conference with a new perspective.

The question repeatedly arose: what actually is the climate crisis? Why is it happening?
In the conversations, we as the organisers all experienced a shortness of breath. Weighed down by the current situation. The magnitude of the crisis and the consequences.

And suddenly the hope arose: how would it be if we were to adopt a different perspective? If a new perspective could give us courage to act? How would it be if we were to take our dismay and look together at what motivates us to act? Young people who have just arrived. Older farmers, experienced and shaped by life. People from the South, North, East and West.
We asked what we as human beings need in order to find our own breath again. Do we need more facts? More head?
Or does it need heart? Poetry? Feeling?

This is how this text came about: from a conversation between the authors and many other people on the occasion of a conference. It is intended to be an appeal. An appeal to adopt a new perspective.

We are attempting something new: we invite you to join the conversation. It will not work without you. It is possible that the text makes demands. That you begin your own process. Ask yourself questions. Think further. Also with your heart.

Will you join in?

The conference *Breathing with the Climate Crisis* took place in February 2021, as a collaboration between the Youth Section and the Section for Agriculture at the Goetheanum in Dornach, Switzerland. It brought together over 1,200 people from 63 countries.
The conference contributions and further information is available at:
www.agriculture-conference.org/de/2021

The Goetheanum is the place where Rudolf Steiner (1861–1925) lived and worked. The Goetheanum is now a school, society and artistic stage, supported by 42,000 members of the

General Anthroposophical Society in over 78 countries.

There are eleven specialist sections where work and development take place in a wide range of anthroposophical fields. Besides biodynamic agriculture, anthroposophical medicine and Steiner-Waldorf education, the sections wish to make a contribution to the questions of our time and human existence.

Short biographies of the authors

Ueli Hurter has been head of the Section for Agriculture at the Goetheanum since 2010 and in the Executive Council of the General Anthroposophical Society since 2020. Biodynamic farmer on the Ferme-Fromagerie de L'Aubier and in the management of L'Aubier up to 2020.

Lin Bautze has been working on the topic of climate change since 2008. She studied Global Change Management (M.Sc.) and Environment and Resource Management (B.Sc.). Practical experience in (scientific) cooperation with and on organic farms. Working for the Section for Agriculture as a project leader for Living Farms (www.livingfarms.net).

Johannes Kronenberg has worked at the Goetheanum, Dornach, since 2019 in research, organisational development, youth work, and on the environmental and sustainability issues affecting the Goetheanum. He has a background in art (B.A.) as well as in Sustainable Development and Management (M.Sc.). He is connected to various foundations and organisations dedicated to change in civil society for a sustainable world.

Publishing information

English Edition *Breathing with the Climate Crisis*
© Hawthorn Press 2023
Published by Hawthorn Press, Hawthorn House,
1 Lansdown Lane, Stroud, Gloucestershire, GL5 1BJ, UK
Tel: 0044 (1453) 757040
E-mail: info@hawthornpress.com
www.hawthornpress.com

Lin Bautze, Ueli Hurter, Johannes Kronenberg are hereby identified as the authors of this work in accordance with section 77 of the Copyright, Designs and Patent Act, 1988. They assert and give notice of their moral right under this Act.

Design and drawings © Mark Schalken, Amsterdam
Edited by Anna Storchenegger
Translation by Lynda Hepburn
Printed by Druckerei Lokay, Reinheim

First published by Verlag am Goetheanum,
under the title *Atmen mit der Klimakrise*
© Lin Bautze, Ueli Hurter, Johannes Kronenberg
www.goetheanum-verlag.ch
ISBN: 978-3-7235-1713-0
©2022 by Section for Agriculture and Youth, School of Spiritual Science, Goetheanum, CH-4143 Dornach

All rights reserved. No part of this book may be reproduced, stored in a retrieval system or transmitted in any form by any means (electronic or mechanical, through reprography, digital transmission, recording or otherwise) without prior written permission of the publisher.

British Library Cataloguing in Publication Data applied for

ISBN: 978-1-912480-87-6

Highest eco-effectiveness
Cradle to Cradle Certified®
print products by Lokay

OTHER BOOKS FROM HAWTHORN PRESS

Designing Regenerative Food Systems
And why we need them now
Marina O'Connell
A unique toolkit of six resilient food production systems: Agroecology, Organic, Biodynamic, Agroforestry, Regenerative and Permaculture. Backed up by striking impact research at the Apricot Centre, Huxhams Cross Farm in Devon, UK
978-1-912480-54-8; 228pp; 234 x 156mm

Gardening for Life
The Biodynamic Way
Maria Thun
This book offers advice and tips on permaculture and growing organic produce, including favorable planting times, building soil fertility and biodynamic preparations.
Fully illustrated in colour, with instructions, index and resources.
978-1-869890-32-2; 128pp; 212 x 160mm

Ordering Books
For UK orders go to our website www.hawthornpress.com or our distributor BookSource:
50 Cambuslang Road, Glasgow, G32 8NB
Email: orders@booksource.net
For USA: Steiner Books: www.steinerbooks.org
Email: service@steinerbooks.org